Brimming with creative inspiration, how-to projects, and useful information to enrich your everyday life, quarto.com is a favorite destination for those pursuing their interests and passions.

First published in 2022 by White Lion Publishing, an imprint of The Quarto Group.
The Old Brewery, 6 Blundell Street
London, N7 9BH,
United Kingdom
T (0)20 7700 6700
www.quarto.com

A catalogue record for this book is available from the British Library.

ISBN 978-0-7112-6594-3
Ebook ISBN 978-0-7112-6596-7

10 9 8 7 6 5 4 3 2 1

Canine expert review by David Alderton

Printed in China

To all the trainers I've loved and learned from. Thank you, Niki, Andrea, Joanne, Cassandra, Mike, and Misa, for showing me and my mom the good boy ways. We are forever grateful.

Xoxo Sproutie

To all the pups I've loved and lost. Thank you, Miss Holly GoLightly, Rugby, Eloise, Bailey, Ace, Chandler, Elwood, Fiona, and Chloe, for watching over Sproutie and me. You are forever remembered as the good boys and good girls you were.

Love, Sigrid

THE GOOD BOY'S GUIDE TO
BEING GOOD

MASTER YOUR HUMANS AND LIVE
YOUR BEST PUPPIN' LIFE

BY BRUSSELS SPROUT

WITH HUMAN TRANSLATION
BY SIGRID NEILSON

WHITE LION
PUBLISHING

CONTENTS

FOREWORD

A lot has changed in ten years of being a full-time dog trainer, but one thing remains constant: the thrill of seeing a dog and their human transform into a team. That moment when they're speaking the same language, and having a blast doing it, is unlike anything else.

I started training Sprout and Sigrid in dog agility in 2017 and have loved watching them grow together. As Sproutie learned to sail over jumps, shoot through tunnels and run over the 5-foot (1.5-meter) high "A-Frame" obstacle, I saw Sigrid's dedication and support, learned Sprout is a thinker and, well, a sassy little smarty-pants! I'm not surprised he's written a book.

One of the most important—and sometimes overlooked—aspects of dog training is the human component. I can teach your dog to sit and stay, but what they really need is *you* to actively invest in the relationship by engaging with them.

The Good Boy's Guide to Being Good captures the importance of that relationship with humor, which is inherent when working with dogs. They have minds of their own and opinions about everything—it's up to us to listen to what they're communicating. For example, when Sprout talks about decorating with a little Lassie poster on his crate wall (see page 27), it perfectly captures how to think about your dog's crate: it's their bedroom, not a 'prison'. It becomes a sanctuary, their own space to chill and relax. Training your dog takes time and effort, but Sproutie and Sigrid will inspire you to become an unstoppable team and fall in love with training your dog!

Nicole Levien CPDT-KA, KPA-CTP, APDT, Co-Owner, Canine Sports & Activity Center (caninesportsactivitycenter.com)

YOU'RE WELCOME

Today, as I settled in for my third mid-morning nap, I felt unusually restless. I led a charmed life with a doting human, so what was missing? Suddenly, I recalled the wisdom on a greeting card I once ate and everything was clear: Sharing is Caring. As a highly intelligent yet impossibly cute creature who's mastered the subtle art of humans, I realized I must share my knowledge, empowering all of puppykind to achieve my level of success and sophistication.

I wasn't born a success with humans. (Born sophisticated, yes.) Humans seemed like a lovable yet highly complicated source of snackies. I wanted to please mine, but her rules were perplexing. Pee here, not there. Chew this, not that. Let's play, now hush! Between naps, I studied the human and analyzed which behaviors yielded the most treats. Some days my task felt impossible. Others, I was on the exhilarating brink of discovery. Mostly, I fell back asleep.

Eventually, I figured out what it takes to be a good boy. I've earned countless snackies, diplomas, and blue ribbons bigger than me. By my precisely imprecise calculations, 10.2 million humans have called me a good boy, and 3.3 million said I was perfect. Clearly, my methods work.

With that, I not-so-humbly present to you my findings. Bring this to your human, snuggle in as they read aloud, and start down the path of being the goodest girl or boy you can be.

PLEASE CONFIGURE YOUR DOG

Everyone in the dog park is staring at their phone. No judgment—
I am too, but I wonder if some focus more on tech than their
dog. Here's the thing: dogs require far more configuration than a
smartphone. Puppies are pure joy, but they're just out of the box.
They don't speak your language, let alone know preferred settings.
It takes time to configure a dog, and you'll need some expert
guidance along the way.

It's all worth it. Sprout pushes a tiny shopping cart, identifies
toys by name, and performs countless other tricks. But what makes
me proudest is that, in elevators, Sprout sits by my feet no matter
how many other people, pups, or bags of food are in there. He runs
free at the beach, always returning when called. I can eat pizza
while he chills in my lap. Sproutie definitely has his moments. He's
a notorious toilet paper thief and loves shoving his ball where it gets
stuck beyond reach. But he's an exceptionally good boy.

By Sprout's first birthday, he had four diplomas and I was driving
30 miles each Saturday morning for his training. No, I'm not a Tiger
[Dog] Mom—we just both love learning, and while we wouldn't
be where we are without professional trainers, our tricks now are
self-taught.

What follows is more than Sprout's amusing wisdom. It's a
collection of useful tips to start you down the path of having a
well-behaved dog and to inspire you to keep going. Put your phone
down and start becoming a puppy power user!

A GLOSSARY OF
VERY GOOD WORDS

It is important for us domesticated creatures to understand humans. But their language is boring, so I improved it with some newer and gooder words. Learn and use these often.

Boop: When a human touches your nose with their fingertip while making the high-pitched exclamation, "Boop!" A mildly peculiar, yet endearing way they show affection.

Dogtor: The human with needles who touches you in weird places. That may sound creepy, but don't worry—it's totally above board and very good for you.

Feetsies: Cute slang for your paws. Just hearing the word makes me want to prance.

Fetchies: Game similar to traditional Fetch, but I play with a special ball because I'm too small to carry a regulation-sized ball. If you get bored, try dropping the ball off furniture and barking until a human comes to halp.

Gooder, Goodest: Clearer terms for "more good" and "the most good". Use these instead of wasting time learning human grammar stuff.

Halp: Human assistance and/or intervention. Most effective when requested alongside puppy-dog eyes and a pout.

Metal Pet: That skinny metal device your human strokes and pays attention to all day. But watch out—one time I saw mine talking to other humans *inside* the metal pet. I shudder at the thought of her getting trapped in there.

Puppers: You, an adorable puppy. Or a grown-up dog, because no matter how old you get, you'll always be a puppers.

Rubbies: A gentle little massage your human gives you behind your ears, on your chest, or wherever else you find pleasing.

Snackies: A superior word for treats and constantly on my mind. You get these tasty little morsels for doing sit or something

else good. Learn where in your house the snackies live and check the floor there regularly for crumbs.

Soapy Wets: When your human puts you in the giant water bowl and covers you with bubbles. Much, much worse than the regular wets. Take your anger out on the towel.

Trickies: Silly things you do at a human's request. Abandon preconceived notions that performing on cue is beneath you—trickies are a gateway to snackies.

Wets: When your human makes the sky pour water. Unclear why they do this, particularly on weekends. Turn around and go back inside for snuggles.

Yip!: A sudden exclamation and release of emotions, like when a human steps on your footsie or you drop your ball where you shouldn't have.

CHAPTER 1

HUMAN INTEREST

THAT TIME I PICKED MY HUMAN AND GAVE HER A PRESENT

I have spent many breakfasts debating how I should begin our journey together. See, I find discussing bodily functions to be rather gauche … but what better way to gain your trust than by sharing what turned out to be my most mortifying moment? Pros and cons of telling this tale were discussed with my faithful ball, but soon we drifted into slumber and I forgot where we landed on the matter. However, I took waking with the urge to conduct my business outside as a sign, so here we go.

The moment I saw the human, I knew she was mine. I worried she'd like my brothers—they were twice my size, handsome, and confident, while I looked like a yam with big ears and skinny legs. But I remembered Mama saying I was the goodest no matter how small I was, so I took a deep breath, stepped in front of my brothers, and shoved my little pout toward this human. She glanced at my brothers. I tilted my head and stared, right through her eyes and into her soul. She smiled as she reached toward me and I melted into her arms.

It was bittersweet. I was nervous leaving home and unsure I was ready to train a human, but excited to start my life. So I bravely packed up my blankie and kissed my Mama goodbye.

I started feeling better once we arrived at my new home. The human gave me presents and swaddled me in plush blankets. She even got down on all fours to play like some big awkward faux-dog. It was weird, but also rather sweet that she was trying so hard to make me feel welcomed, so I played along and let her act silly. And when I started to miss being curled up with my siblings I found solace snuggling in the warmth of her armpit.

The next morning after breakfast, the human scooped me up and placed me in a metal enclosure. "Okay, good boy, you hang tight for a bit," she said, patting my head. *Wait, what?* I looked up at her, hoping my big blinking eyes would cause her to melt and stay with me forever. They didn't. She began to close the metal door and with one last pat, said, "I'll be back soon."

Okay. But. Are you *sure* you'll be back? I lacked sufficient data to determine whether this was a credible claim. What if she was lying and this was how it ended? Me, taken from my family, lulled into a false sense of security and left alone never to cuddle again. I felt a pit forming in my tiny pink belly as panic began to swell. There were tears, so many tears, pleas to the heavens, and failed attempts to escape through the cracks of my newfound fate. The pit turned to a grumbling terror deep within. And then … there was poop. Followed by immediate regret.

Oh, yip! What had I done? And what would she think if she *did* come back? I had no good reason to doubt her claim, and this was not an appropriate way to thank her for the snuggles and snackies received to date. Or maybe it was? I mean, she seemed excited when I pooped late last night? *It's okay, stay calm,* I told myself. Myself answered back with panic. I began pacing, frantically trying to gather my thoughts, but I stepped in poop and it began to smear. I know, so gross, but it got on my feetsies and blankets and all over that stupid stuffed frog who continued smiling notwithstanding my crisis. I started pawing at the frog, trying to rub the poop off on him. He just smiled and stared back. I lost it. "WIPE THAT RIDICULOUS GRIN OFF YOUR FACE, YOU FOOL," I barked. I couldn't hold back any longer. More tears. Primal screams to the heavens. And then the door opened. Oh, yip! She was back.

YOU ARE GOOD

The first rule of life is: All. Dogs. Are. Good. Well, except that one who knocked me down at the park, he was a jerk. But if you take one thing away from my findings before drifting into slumber, it's that you were born good and will always be good no matter what else happens. If something goes wrong, don't internalize your emotions and turn to self-blame. It was definitely a human's fault. And on the off chance someone says you're bad, consider it an open invitation to pee on their shoe.

Dogs are incredibly well intentioned and want to please us. Just take one look into your pup's eyes, and you'll see they are filled with so much good-dog it'll make you want to cry. If your pup didn't do what you wanted, recognize it's not because they are "bad," and instead of blaming them, think about how you can be clearer in communicating your expectations.

DON'T PANIC

The second rule of life is: Don't Panic. And I say this as a largely reformed, yet highly skilled, panicker. You're wasting energy you should preserve for power naps or fetchies. Trust that the human will come home. I mean, they *have* to because home is where you are and nowhere is better than where you are. Being alone can be scary, but once when nibbling a self-halp book, I learned that repeating a mantra can soothe anxiety. Try: *Keep calm. Puppers on.*

I get it. You want to spend every moment staring at your pup, for no creature has ever looked so cute licking your toe. But it's best for you both to create boundaries and learn to be apart early on. Even if you don't leave the house all day, practice being "separate while together." Crate your pup (see page 27) while doing housework or leave them in another room as you eat dinner. Getting mail and taking out trash are great ways to practice leaving your dog and showing him that you come back soon.

RECOGNIZE THAT TELLTALE CRINKLE OF THE SNACKIES BAG

There is no greater reward than the joy of making someone else happ—
Oh, who am I kidding. Snackies are everything. By my highly accurate
guesstimations, 23 percent of the time I do something good it is, in fact,
out of the goodness of my heart. However, 9 percent of the time it is
because I would like a snuggle, 6 percent because you have the ball, and
62 percent for snackies. Look, I said I was a good boy, but I never said I
was stupid. I am, however, fairly discerning when it comes to what I eat
so that I can look and feel my best—only the finest of snackies for the
finest of creatures.

Are your back pockets awkwardly filled with little sausage pieces?
No? Then you don't have a puppy. Carry treats with you so you
can reward good behavior and new experiences. Look for treats
that easily break into teeny tiny bites so your pup doesn't overload
on treats. A cooked chicken breast, slice of turkey meat, or little
bite of sweet potato can end up being tastier, healthier, and more
cost-effective than some bags of store-bought treats. Make sure to
check with your vet about what human foods are safe to give your
pup and how many treats they can have while staying healthy.

REMEMBER WHAT
MADE THEM HAPPY

Honestly, a lot of success with humans comes down to mimicking behavior they find pleasant. Is it really that simple? Kinda. Does that mean you have to do things that feel foolish and potentially pointless? Definitely. So when you hear that sing-songy, "AWWW WHO'S A GOOD LIL' PUPPER" come out your human's mouth, note what you were doing and just keep doing that for the rest of your life.

We learned from trainers who use "positive reinforcement" and found the method so understandable. The idea is that it's easiest to teach someone by rewarding them for doing something right, not by scolding them when they do it wrong. This works well with puppies because your attention and love can be just as rewarding as treats. Give them lots of happy praise for doing what you want. If something goes wrong, don't scold or make a fuss—just be neutral and move on.

HOW TO OPEN A BOTTLE OF WINE

Here's how I think about positive reinforcement training, and like all good explanations it involves wine. Imagine needing a clueless friend to open a bottle that has a cork. But … you can't tell them how a corkscrew works, only watch and yell, "nope not that way" and "wrong again," as they try to do it. By the time the bottle is open you're beyond frustrated and ready for a glass.

Now imagine you're that friend. You're not clueless, just used to the convenience of a screw-top. You pick up the bottle and turn it over as you reach for the corkscrew. "NOPE. Not like that," yells your friend. A few minutes later, you've also learned that you don't lift up the corkscrew's arm-thingies to start and you definitely do not stab the cork repeatedly until it breaks. Or so you think? Maybe stabbing is right but you did it at the wrong angle? Whatever, the bottle is still not open and you desperately need a glass of wine after this frustration.

When you reprimand your puppy—let's say for chewing your shoe—you're leaving them wondering, *Uh-ohs, am I not supposed to chew shoes? Are laces okay, but heels off-limits? Or maybe I chewed too close to her, and I should do it in secret next time?* They don't know what exactly they did wrong, only that you're upset. Don't yell. It's your fault for leaving them alone in a walk-in closet. Take the shoe away, give them a safe bone, and tell them what a good puppy they are as they nicely chew what you want them to chew.

HANG OUT IN YOUR ROOM

Being left in a crate may sound like some sort of Dark Ages treatment, but it's actually the goodest. The whole house is my kingdom, but my crate is extra special because it's the one place the human can't go. A little safe palace. I take most of my daytime naps there, and can rest easy because I gathered enough data to trust the human will return after putting me in my crate. So, invest in cozy bedding, hang up that poster of Lassie, and retreat to the comfort of your very own bedroom.

Crates are great for housebreaking, but also provide peace of mind. When I leave Sprout out in the apartment, he sits at the door staring anxiously until I return, and I worry if I accidentally left something out he can get to. But in his crate he sleeps stress-free and I know he's safe with water, room to move, blankets, and his "beloved" stuffed frog. Learn how to crate properly: help your pup adjust gradually, remove collars or choking-hazard toys before leaving, , don't leave them longer than they can "hold it," and treat it like a bedroom not a prison cell.

SPEAK UP FOR YOURSELF

As perfect as I am, I must admit I can't speak words like a human—yet. Until then, us puppers must find other ways to talk to humans. For example, a tucked tail and sad puppy-dog eyes says, "Nopes, this is scary," while a polite sit paired with a head tilt is scientifically proven to yield 4.7 times more snackies than barking. Oh, and don't forget to use your native tongue. Like, literally—just lick their knee repeatedly to say, "I love you."

Don't overlook what your pup is trying to tell you. Watch and understand their body language—did you know yawning and licking their nose are sometimes signs of stress? Learn and honor your pup's preferences—where do they like to be scratched most? And try learning a few words of your pup's language. Dogs often initiate play by bowing to each other, so try getting on all fours and play-bow with your pup the next time you're ready to play!

CLEAR RULES, CAN'T LOSE

Everyone knows the worst thing about humans is that whole fake ball-throw routine, but a close second is their confusing rules. Like, how come you gave me a bite of steak precisely seven weeks and three days ago, but tonight I can't have one? What do you mean "special occasion"—isn't every night with me special? To understand their rules, you must track which behaviors got rewarded and which did not, until you spot patterns. It's challenging, but once you figure out the *actual* rule, life gets much easier.

Keep your rules simple and consistent. Don't make exceptions—your dog won't understand the nuances. Prada and Payless look equally delicious, so don't let your pup chew an old shoe and expect them to know the designer ones are off-limits. And make sure everyone in the house is aligned on the rules and their importance. If you don't let them on the sofa, but your girlfriend does, then what is their lil' puppy brain going to think when both you and the girlfriend are sitting on the sofa?

BE A CREATURE OF
HABIT (AND LEISURE)

I'm all for surprise trips to the beach but, ultimately, I prefer predictability over spontaneity. Each day starts with the human lovingly coaxing me out from under the covers. She carries me on a walk, makes my breakfast, and leaves me to enjoy peaceful slumber, later returning so that she can make my dinner, play fetchies, and snuggle. I will get a case of the so-called "zoomies" from 11.30–11.32 p.m., wherein I run amok in little circles until I collapse into a pile of puppy dreams, tucked under the covers until my perfect day starts all over again.

Dogs thrive with routine, and consistency helps them adjust to your rules. By managing your pup's schedule, you set them up for success. If your pup needs a potty break every four hours, don't let them play unsupervised during hour three as they're more likely to have an accident. Make sure to form routines that actually work with your non-dog life—don't feed them dinner at 5.30 p.m. sharp if half the time you won't be home from work until much later.

FIND YOUR SUNNY SPOT

Living with humans is an emotional roller coaster. They will annoy you, refusing to let you eat mystery goo off the sidewalk. They will soothe you with gentle little belly-rubbies and lullabies. One minute, life is so good it feels like a movie scene with you and your human running through green fields, laughing and smiling in slow motion. But then suddenly you're questioning your sanity, wondering if you were good enough. (You were.)

On days when you feel incredibly whelmed at trying to figure out your human, remind yourself why you love them. Not just the obvious reasons like food and the warmth of their knee-nook, but the special little gestures that tug at your puppy heartstrings. For me, it's how my human gives me little chin-scratchies when she catches me staring up at her. She often gives me snackies without even making me sit. And she periodically stops whatever she is doing and repositions my bed so I stay napping in the sun. I initially assumed this was a standard human duty, but now I believe it is her way of saying she cares for me most—and when she does it, I feel warm inside and out.

So go on, find your own metaphorical sunny spot and carry it with you always.

CHAPTER 2

HOUSE RULES

THAT TIME SHE MADE IT WET

I awoke to the sound of the human's weird cooing voice as she lifted me from the plushness of my bed and gently swaddled me in bright orange fabric. She carried me outside per standard protocol, but as she set me down I noticed something unpleasant, so *very* unpleasant. It was wet. Everywhere. Ew. I tried to stand without putting my feetsies down and looked up in disgust. Was she serious? She nodded and pointed, her stupidly cheerful "Go potty" echoing in my ears. Double ew.

I forced myself to walk, thinking perhaps the wets didn't exist by the next tree, but a few steps later I hit the end of my leash. I tugged, to no avail. The human shook her head and pointed, her infuriating "Right here's fine. Go potty!" taunting my ears. This is so not pleasant. I squatted and peered up. Wetness pelted against my pout, washing away all my dignity. Why would she make it wet? Was she joking or would it be this way forevermore? I stared up at her as hard as I could, hoping my look of shame would be forever engrained deep in her soul.

But then she smiled and knelt down with a snackie, her sing-songy voice calling me a good boy. My tiny puppy insides warmed with pride. I was a good boy, the goodest of them all. She scooped me up, shielding me from the wets as we ran back inside. Maybe we'd make it after all, just me and her conquering the world. The worst was over. Or so I thought.

The human scratched my little chin and unclipped my leash. As she began fiddling with the bright orange swaddling, I drifted into thoughts of breakfast and the day's agenda. Perhaps after my second nap I'd work on unraveling the rug or, better yet, sneak attack that fuzzy little squirrel. I was imagining it squeaking in fear at my ferociousness, when all of a sudden KKKKXRRIIIIIIICCHCHCCHCCHHHHHH!!!!!

Oh, yip! It was the deafening roar of the entire universe being ripped apart, right beneath my chin. There was no time to think. I sprung from her arms and took cover underneath the table, unsure whether my trembling was fear or the reverberating aftershock of the end of the world. What was happening? Was my human still there? Was I even still in one piece? It was too scary to look.

An eternity later, I gathered my puppy courage and peered out. The human was there, sitting on the floor, my bright orange swaddling limp in her arms. Puddles of guilt seemed to well in her eyes. "I'm sorry, puppykins. It's, it's just, it's called Velcro®, Sproutie honey." KKKKXrrriiiiicchcchCCHHHH!!!, the fabric went as she ripped it apart. I flinched, but saw a snackie dispensed from her hand. A peace offering? I leaned out to grab the morsel, quickly darting back to safety. Another Kkkkxrrriiiiicchcchccchhh! and another snackie dispensed. I crept out from under the table. She smiled. I pouted less. Perhaps we'd make it after all. Just me and her, and maybe some snackies.

YES, YOU DO THAT OUT THERE

This whole doing your business thing may feel undignified, perhaps even mildly shameful. And don't even start me on the awkwardness of a staring, cheering audience. However, consider your alternatives: Soiled bedding? Stenches wafting past and interrupting your sweet dreams? Much worse. That being said, it is your human's duty to provide adequate access to the outside. If they fail, try your best to hold it, but ultimately do what you need to do. I recommend relieving yourself in the tiled room with the giant water bowl—you know, where soapy wets happen. Humans *love* a good twist of irony.

Do you always do things perfectly on the first try? Doubtful, so cut your pup some slack and recognize that accidents happen and may even be your fault. Housebreaking comes down to managing your puppy so they're in the right place at the right time. Avoid letting them have free rein of your home when they're in a "code yellow," meaning enough time has passed since the last potty break that they're at risk of going at any moment.

YOU'LL NEVER FIND THE PERFECT SPOT

Look, not everything in the world is squeakers and rainbows. No matter how many times you circle or sniff or tell yourself that the next tree is better, you will never find just the right spot. Trust me, I've tried. Ultimately, it's best to accept this reality with grace, do your business wherever, and save your energy for exploring better things. Seriously, though: would you rather tug on your leash to get to yet another tree … or to sniff a half-eaten hot dog some dude dropped on the sidewalk?

The idea is your pup shouldn't even know wandering for a spot is an option. Pick a place for them and stick to it. Let your pup wander to the leash's end, but keep your feet planted even if they tug to go farther. Similarly, save the actual walk and exploratory sniffing as a reward for after your pup has gone to the toilet. Otherwise, your pup may develop a habit of needing to wander before going, and I doubt you'll enjoy the quest for the perfect tree when you're running late for work.

WHEN IN DOUBT, TRY A POLITE SIT

This is a somewhat controversial opinion, as many dogs find barking to be an enjoyable pastime, but I consider the behavior to be rather uncivilized. Granted, barking can communicate necessary alerts such as your ball rolling out of reach, and we all let out a yip or two here and there, but otherwise, barking does not reflect the demeanor of a mature and sophisticated good dog. More importantly, it's not effective: according to my scientifically proven estimates, you are 16.4 times more likely to get what you want by sitting quietly and staring up at a human, instead of barking at them. Bonus point for head tilts.

Teenage Sprout started "demand" barking—communicating his wants by barking incessantly. Trust me when I say you'll want to break this annoying habit quickly. Fortunately, positive reinforcement works here too. Teach your pup to ask for things nicely by sitting or lying down. Practice regularly: before opening the door, putting down the dinner bowl, or throwing the ball. Ignore demand barking, but the moment your pup offers a nice sit, give them what they want. Pretty soon, your pup will realize manners are the key to everything.

GET IN THE BIG BED

Do whatever it takes to get yourself in the human bed. Here's the secret: humans have this "let sleeping dogs lie" law, which means if you fall asleep right in the middle of the bed, on their head, or wherever else you find comfy, the human is *literally not allowed* to wake or move you and must stay in whatever awkward position they're stuck in. Don't worry, this isn't selfish of you. I mean, what could a human want more than to spend hours awake just watching you dream sweet puppy dreams?

I have to disagree with Sprout here. Whether or not your dog sleeps in bed with you is up to you, not your dog. Think long and hard about this decision, because once your pup spends even one night in your bed it may be hard to get them to stop. This is going to be one of the last boundaries you want to blur, so try having your pup sleep in their own crate or bed for the first year, and then decide what makes sense for you and your household.

THE ART OF NEGOTIATING
WITH A PUPPY

Puppies are tiny bundles of softness, joy, and temptation. One look into their eyes and you'll do anything. You'll bail on your friend's sister's 34th-birthday blowout because your dog wanted to play tug (okay, maybe that's an easy choice). You'll call in sick to work because your puppy looks sad that you're leaving. (Dear boss: I swear I haven't done this.)

Whether it's napping as they snuggle under your chin, frolicking all afternoon in the park, or recording them attacking a pillowcase, you should absolutely enjoy every moment of puppyhood. But be mindful that what you do now creates lifelong habits. Your puppy running off with every item as you unpack groceries may seem cute, but in three years' time do you really want to chase a grocery-thieving dog after a really long day?

Rules are particularly important at home. Everyone should feel relaxed in their home, and it can really strain relationships if you and your dog are stressing each other out or if you and your partner aren't aligned on how to raise a pup.

Commit to consistently enforcing rules for the first year. If something is important enough to even be a rule, then it's worth the added energy to enforce it. After a year, a few things happen: you either no longer consciously enforce the rule because it's now second nature; you realize the rule still matters (which motivates you to keep it); or you realize you don't actually care about the rule (which allows you to intentionally relax it). But ultimately, the decision should be made by *you*, not those perfect puppy-dog eyes.

LEARN TO ENTERTAIN YOURSELF

Life can be boring, like when the human makes the sky do the wets or spends all day with the metal pet in her lap. Learn to entertain yourself. One easy way is to grab the end of the toilet paper roll and see how far you can run with it. Weave between chair legs for bonus points. Alternatively, settle down with your favorite bone for a few hours of peaceful chewing. And yes, I just suggested you cure boredom with a seemingly mundane activity, but there is just something so soothing about chewing a bone.

I don't love Sprout's toilet paper advice, but he's right: if you don't teach your pup to entertain themselves, they'll find their own form of entertainment. And odds are that involves them getting up to something they shouldn't. Avoid issues by ensuring your pup has a puppy-safe chew toy or food-stuffed bone at all times, but make sure you don't leave them alone with a potential choking hazard. Chewing not only keeps your pup busy, but also helps soothe anxiety. And, of course, it'll save your favorite shoes from an untimely demise.

BARKING AT THE DOOR
IS UNNECESSARY

Doorbells are one of the most commonly misunderstood aspects of human households. I mean, *even I* got this one wrong. I totally thought I was halping by alerting my human that another human had arrived at my house. Well, turns out that ringy sound is the alert. By barking at the door you're telling a human what they already know, which is simply a waste of your time and energy, not to mention mildly tacky. Instead, sit there nicely so that whatever human enters is humbled by your good-boy presence.

I don't know how to get your dog not to bark at the doorbell. Seriously. Teaching what not to do is a time-consuming, frustrating approach. Flip things around using positive reinforcement and show your dog how they should act when the bell rings. Is it lie down wherever they are? Go to bed and sit nicely? Decide what routine makes sense for your house, and remember, a bark or two is natural—your goal is to avoid excessive barking or freak-outs.

IF YOU WANT HUMAN FOOD, FIND THE NEAREST SUCKER

My meals are tasty and nutritious and all that stuff, but there's something so exhilarating about the food the human eats. Unfortunately, her one flaw is that she won't share it. Never ever. I used to get jealous seeing humans on TV sneaking their dog bits of yumminess underneath the table, but then I got smart. It indicated that some humans *can* be manipulated into sharing. Now, I focus my energy on convincing other humans to share bites with adorable me. Most don't, but nevertheless I persist.

I can eat almost anything with Sprout in my lap because I've literally never fed him off my plate. Someone, who shall remain nameless, fed baby-Sprout part of a ham sandwich and was forever stuck with Sprout going crazy whenever they ate. But, far more important than enjoying your meal in peace and quiet is the fact that many people foods, even 'healthy' items, can be dangerous to dogs. It's best to just avoid feeding your puppy off your plate. If something is really share-worthy—and safe for your dog—save a tiny bite for once you finish eating, walk away from the table, make your pup sit, and reward them with the bite.

MAKE FRIENDS WITH THE VACUUM MONSTER

I overheard some Yorkies at the park talking about the monster at home. One said he hid from it, and the other called him a weenie and said she just barked to scare it. And I was like, "Wait, what? You mean the treat dispenser that eats human shedding from the floor? He's the goodest!" They looked at me like I had five tails, but it's true—I became friends with vacuum monster early on, and I still eagerly greet him when he shows up (which isn't often enough, if you ask me). A cautionary tale, though: vacuums are cool and stuff, but they can bite, so stay away from their mouths.

Think about what could be loud or scary in your home—vacuums, blow dryers, etc.—and gradually introduce them. For example, try turning the vacuum on low, across the room, while giving your pup treats. Once they're relaxed, bring the vacuum closer while still treating. Keep repeating until your pup is right next to the vacuum (but away from any suction—safety first!). Sprout would even eat treats off the vacuum while running, and years later he still comes running whenever he hears the vacuum!

AN ODE TO RULES

Dear rules: I once thought you were beneath me
But a good boy I very much wanted to be
I stopped eating the rug and swallowed my pride
And by you, silly rules, I began to abide.

At first it was hard, learning to sit and say please
But I found inner strength (and a love of snackies)
I conducted business outside and learned shoes were no toy
And that's when I knew I was a very good boy.

I think of my puppyhood now with great fondness
Granted, it may be because I'm just flawless
But, dear rules, I believe you may have played a part
So, thank you, from the bottom of my teeny tiny heart.

CHAPTER 3

BRAVE NEW WORLD

THAT TIME I FELL IN LOVE
WITH THE WINE STORE

I've always been perfectly perfect, but once upon a time I was terrified of shopping. I mean, I wasn't opposed to the *concept* of shopping. In fact, I quite liked the human buying me things. But entering a store was scary. What did it contain? Nail cutters? Monsters? Snuggles? I was not risking my life to find out.

The human, however, had other ideas. She loooooved shopping. Like, not just buying me things, but also the act of walking into a store. I was awestruck by her fearlessness in going through doorways. Such confidence, such grace. I felt inspired and vowed to conquer my fears.

I cautiously peered into my first doorway, and when I didn't see monsters I took a deep breath and moved my bravest little paw. A snackie was bestowed upon me. This was promising. Another step, another snackie. Half-a-pup inside and still alive. The human smiled, and said, "You got this lil' pupperkins!" I reminded myself how bravely she did it, squeezed my eyes shut, and went for it. All 3.2 pounds of me leaped inside. Yippee!! There were cheers. Snackies! Chin-scratchies while she told me I was the goodest! I nearly blushed at all the fuss she made.

A few weeks later on our evening walkie, the human stopped at a new store. She opened the door and I peered in. It seemed safe, albeit boring and clearly lacking things for her to buy me. I looked up. Really? This is more important than the dropped box of chicken wings back at the corner? Puppy-dog eyes rolled, but I entered nonetheless. The store was filled with old bottles. They smelled like bottles, but old. I sat there while she stared at the bottles. Good grief lady, just pick one. They. All. Smell. Boring.

Eventually she picked one and made her way to the front of the store to engage in humanly interactions. The store human smiled and gestured at me, reaching for a box of unknown contents. Oh no, was it nail cutters? The end of the world? Had I been lured into a false sense of security by the boringness of old bottles? But as the store human removed the lid I caught the heavenly scent of snackies and my little tail began to waggle.

The store human bent down, broke the snackie into tiny bites, and held one out. I looked up at my human, remembering cautionary tales of taking food from strangers. She smiled and nodded. I leaned forward and plucked the tiny snackie from his hand. He held out another tiny snackie and another and another. I couldn't believe it. How did I not realize stores have snackies … FOR FREE? This was better than gifts.

"Okay Sproutie, that's enough," she said, thanking the store human and turning to leave. *Wait, what?* I firmly planted my puppy-butt. "Puppers, let's go now," she said with a gentle tug. I laid down, shot her my best side-eye, and tilted my head lovingly at the store human. He and his box of snackies were my world now. But my human scooped me up and carried me out. I looked over her shoulder, staring longingly at the store human. *Don't you worry, we'll be back.*

HOUSE RULES APPLY OUTSIDE TOO

Humans believe they control things inside *and outside* your home. Ummmm, okay, but like where do I run the show? No seriously, where? In my tiny bedroom where you can't fit? Because that seems unbalanced. At any rate, set aside your pride and listen to your control-freak humans no matter where you are. It's not always easy with all the garbage and new BFFs you'll encounter, but it's part of being a good pupper.

Dogs can be contextual, so don't stress if your pup sits perfectly at home but ignores your command outside. People tend to train in the same quiet spot at home, but you'll want to also practice in new places and circumstances. Once you've mastered a new command at home, progress to a distraction-free spot outside your home, like a patio or apartment hallway. Work your way up to bustling sidewalks and parks. Long leashes (20 feet/ 6 meters) are great to safely practice "stay" and "come" outside.

BE OPEN-MINDED

My experience with the old bottles store taught me that life is about perspective. Instead of assuming every doorway led to doom, I began imagining everywhere led to free snackies. Granted, after a few weeks, I determined that only 18 percent of doorways actually led to snackies, but the *possibility* made each doorway worth it. Apply this positive outlook to all new things you encounter and explore the world with an open mind. Sure, there will be a lot of boring human things, but I bet you'll find a few free snackies along the way.

Your puppy is a tiny sponge, ready to soak up the world. Dogs are often more receptive to new experiences earlier in life, so give your pup plenty of opportunities to explore! Come up with a puppy bucket list of all the different types of people, places, and things you want to introduce them to by their first birthday. Respect your pup's comfort level when tackling your list and use treats and praise to help create a positive experience.

TAKE IT DAY BY DAY

We all know a pup's gotta learn to walk before he can prance, and that same principle applies to navigating the outside world. If you're a sensitive soul like me, this may take time. Remember to keep your mind open, but if something new starts to feel overwhelming just let your human know so they can slow down and halp you adjust. And once you feel more comfortable, you can start prancing through life.

The goal is to help your pup grow by gradually upping their comfort level. Do not throw them into a scary situation and hope they adjust. For example, if you want to introduce your pup to bicycles, let them approach and sniff a still, riderless bike. Once they're relaxed, get your pup used to someone walking a bike past them, and keep progressing (and treating) until someone can ride past without your pup caring. If your dog really struggles with fear, consider having a professional trainer help you safely address the issue before it becomes a lasting concern.

ODDS ARE, THE WORLD ISN'T ENDING

I don't mean to scare you, but things outside get pretty real. There are booms and bangs and sirens all at once, like some horrible soundtrack to the end of the world. Statistically speaking, however, it's probably not, but I get it—in a moment of terror, it's hard to recall statistics. Take a few deep breaths. Scary noises usually end as quickly as they started. Plus, they often result in human sympathy and free snackies, which is always awesome.

No matter how hard you try to help your pup gradually adjust to outside noises, you can't avoid unexpected noises like a motorcycle revving up 3 feet (1 meter) away. Be ready to counteract a startling experience with something positive. Try bringing treats with you whenever you head outside for extra reinforcement, although your cheerful voice and reassuring praise will also help soothe your pup. Important side note: hand your puppy treats outside, instead of dropping them on the floor—you don't want them to think it's ever okay to eat something off the ground.

BE A [DOG] MAMA BEAR

The outside environment is often beyond your control. Experiencing new things and people can be great for socializing your puppy, but as hard as you try it may not all be positive. Strangers will pick your pup up without asking. Others will put their hand over your pup's tiny head to give them head-noogies. And while I hope this never happens to you, someone on the subway just may try to bite your dog's ear (it happened to us). Everything happens in an instant so you may not be able to prevent it, but you can put an end to it and start to manage how people interact with your dog in the future.

Your pup sets the rules of engagement and you are their enforcer. Be polite but firm—you don't owe strangers anything, but you do owe your dog positive, safe experiences. If strangers holding your pup makes them nervous, don't let strangers hold them. If your dog doesn't like being petted on the head and someone bends over to pet them, say, "Oh hey, give him a lil' butt-scratchie please, he doesn't love strangers near his head." Nobody wants to scare your puppy, so guidance is usually well received.

Same goes for managing your pup's interactions with other dogs. Humans should communicate with each other before their dogs do. Never assume another dog is friendly just because yours is. Ask before letting your exuberant puppy run up to another dog. If you know your dog is shy, tell other owners so they can manage their own dog. If you know your dog doesn't like meeting other dogs leashed on the street, you need to speak up and *avoid* the interaction. There's no requirement that dogs must sniff each other as they pass.

FIND YOUR CANINE CREW

Playing with humans is great—even tiny humans can throw balls farther than any dog I've met, and most humans don't slobber on toys. That being said, you're a pupper and should have pupper friends. If you're outgoing to begin with, congrats, you probably already have fifty zillion friends. But if you're shy at first, like me, don't worry—you got this. It's okay to start with just a few sniffs and take it slowly. If needed, you can gently let another pupper know if he's coming on way too strong. Gotta respect the introvert, man. P.S. That's my girlfriend Penny pictured, isn't she perfect?!

Give your puppy opportunities to safely socialize with other puppies—trainer-hosted playgroups can be great for supervised play. Not all dogs like to play-wrestle, so don't worry if yours is shy at first (or forever). Let them socialize at their own pace. Sprout hid under a chair and snapped at another puppy who jumped on him. I was horrified, but the trainer explained that's how dogs communicate. Sprout learning to express his need for space helped the other puppy learn to read social cues and play gently. Sprout soon started playing with others and while he never became a wrestler, he's made plenty of friends along the way.

YOUR DOGTOR IS
THERE TO HALP

My first date with the dogtor started off great. Good sniffs, and she seemed nice and smart, but suddenly it was like, "WHOA, hey lady we just met … how 'bout some snackies before you grab that?!" It got worse when she pulled out a needle. I vowed to never see her again, but we kept crossing paths. Eventually, I realized she cared in her own weird way and wanted to halp me feel my best. So I put my faith in her and never looked back. And guess what? I look and feel like a million snacks.

To help your pup feel safe and comfortable at the vet, practice playing "doctor" at home by touching and inspecting your puppy all over: look in ears, open mouths, pick up paws, poke and prod, giving treats and praise as you go. Bring extra treats to the vet's office and let your pup sniff around and get familiar with the space. Also, talk to your vet about pet insurance—most policies don't cover preexisting conditions, so it's best to get coverage early on. And please, please, please take a pet first aid and CPR class.

TINY HUMANS ARE
VENDING MACHINES

I was skeptical of tiny humans at first. I read they were 2.8 times more likely to play than normal-sized humans, but they struck me as somewhat unruly and rather unpredictable. And then I discovered something important: they can dispense snackies. Every tiny human I met would present me with a snackie, and then get all giggly when I nuzzled it out of their hand. Like, how pure and cute is that? Tiny humans also accidentally drop food a lot and can be easily manipulated into sharing with you. All of this is to say, give tiny humans a chance.

Even if you don't have kids, consider how to safely introduce your puppy to them. Whenever we met a kid out on our walks, I'd have them give Sprout a treat. Both loved it, and Sprout started approaching kids like vending machines. However, even obedient children and puppies shouldn't interact without close adult supervision—neither really understanding boundaries, and things can go from good to bad in an instant. Manage the interaction as well—Sprout isn't snuggly with kids, so I have kids play fetch or do tricks with him so they both have fun without close contact.

SNUG UP IN A SWEATER

You know that feeling when your human wraps you in a soft blanket like a snug little puppy-burrito? It's cozy, but you can't move. Sweaters, however, are basically puppy-burritos with feetsie holes. They help you stay warm *and* make you look extra adorable, meaning you're more likely to get good-boy compliments from humans. But hey, not all clothes feel good—they can be *too* snug or touch your tail (my pet peeve). If you don't like your outfit, refuse to move until a human takes it off.

You have to respect your dog when it comes to clothes. Sprout is fine wearing them—he actually dresses himself, lifting his front legs as I say, "One paw, two paw." But don't force it if your pup is uncomfortable—there are countless other options out there, like cute collars, dog tags, and bandanas. If it's cold and a coat is essential, make sure it fits well, and help your pup adjust by practicing putting and keeping the coat on, giving treats along the way.

DREAM. BELIEVE. ACHIEVE.

It wasn't just doorways that scared me as an itty-bitty creature. Towels, puppies, motorcycles, wind, you name it and I'd hide behind my human's ankles trying to remember my deep-breathing exercises, eventually finding the courage to creep out and try again. Now, every morning I rise from my exceedingly plush bed like the shining star I am, knowing there is nothing in life I cannot achieve. Well, except perhaps reaching that bag of unknown contents left high upon the countertop, but whatever. Now I have confidence.

How did I do it, you ask? Well, I would like to thank my human, the most patient, supportive, and beautifulest creature to ever exist. She showed me that Velcro® couldn't fight back, she told strangers to give me butt-scratchies, and she made snackies fall from the heavens when scary noises boomed like thunder. She believed in me before I believed in myself. And now I believe in you. So go on, have the confidence to believe that one day you will get that bag of unknown contents.

CHAPTER 4

TEACHER'S PET

THAT TIME I LEARNED THE EASIEST TRICKIE EVER

I began to formally study the theory and practice of Good Dogism when I was just fourteen weeks old. I was quite nervous, but also optimistic about what my future held. I heard that many of New York City's finest good dogs had graduated under my same headmaster. And at this point I was starting to accept that I may not grow up to be a tall and ruggedly handsome dog. So perhaps I could instead become a small and absurdly smart (and handsome) dog.

We arrived at my school and I marched confidently through the doorway and toward life as a scholar. Ten feet later, however, I was met with an insurmountable challenge: stairs. I stood for a moment, wide-eyed, trying not to panic. But then some goldendoodle knocked past me and bounded up the stairs. *Nope, nope, nope, I can't.* Panic took over and I turned to run. The human gently caught me and picked me up, looking me straight in my eye. "Sproutiepie, no one expects you to go up a stair twice as high as you. And are you really going to let that doodle think he's better than you?" She gave me a chin-scratchie and marched up the stairs confidently enough for the both of us.

There were four other puppies in class and they were all bigger and older. I did not sit in the front row. I sat behind my human's foot and cautiously peered out from time to time. There was a lot of theory at first. The human appeared to listen and take notes, but I was bored out of my puppymind. I thought there'd be snackies?

Eventually, she put down the notebook and turned toward me, pulling a tiny little morsel out from her pocket. *Finally!* I took a step out from under the chair. She held the treat right in front of my nose, and as soon as I leaned in for it, she started moving it away toward the ground.

Wait! *Wait for meeee*! I leaned in and followed the little morsel until I was lying on the ground. "Good!" she cheered and gave me the snackie. Okay? I mean it was delicious and all, but that whole process seemed unnecessary. She reached for her pocket again, and I sat back up eagerly, my tail waggling. I licked my pouty little lip as I leaned in for the morsel, but once again she pulled it down away from me. My pout followed closely and a split-second later I was lying down again. The snackie was mine.

This kept happening. To be honest, it felt really rude of her, but after like ten times I had a thought: was this whole snackie luring thing an attempt to make me lie down? Seriously? Did she forget who she was talking to? I love sleep more than anything, do I really need to be *taught to lie down*? She'd give me a snackie in exchange for acting leisurely? Lolllll, sure lady. She gestured her hand again and I laid down with the most enthusiasm ever known to puppykind.

It was in that moment that I embraced my intellectualism and found my purpose—doing simple things in exchange for snackies. I mean, I'd need to negotiate my pay, as I fundamentally disagreed with the notion that my actions were worth only one morsel. But just like that, being tall, bark, and handsome was no longer on my mind. From here on out, it was all about the good-boy.

TAKE YOUR HUMAN TO SCHOOL

Taking your human to training is one of the best things you can do. I was skeptical at first, but as soon as I realized our teacher could speak both pupper *and* human, it felt like a weight was lifted off my tiny puppy shoulders. Thank DOG, finally someone could halp my poor human figure out what she was doing. Having to help your human practice and stuff is definitely a commitment, but they'll get so much out of training, and having a well-trained human ultimately makes your life easier.

I could gush about dog training all day because there are so many reasons to take your pup to school. A good puppy class goes beyond "sit" and "stay," covering issues like housetraining and socialization in detail. You'll bond with your puppy and earn their trust. You'll work with a pro who can answer questions and help you troubleshoot. Read about the course and instructor before signing up to make sure it sounds like the right fit—we love classes that use positive reinforcement methods.

MINIMAL PRACTICE
MAKES PERFECT

At first I really worried my scholarly pursuits would interfere with naptime. Would I be drilling sit-stay for seventeen hours straight? Because eww, no thank you. But I was pleasantly surprised to learn homework wasn't all that involved—just a few minutes before the human gave me dinner. We also practiced in elevators and waiting in store lines, which was fine because I was already awake and bored. Anyways, don't stress because being good is easier than you think.

Training your dog is a commitment, but it shouldn't be overwhelming. You don't need hours at a time, and it's actually better to practice in short spurts to avoid your puppy (or you) getting bored, tired, or frustrated. Pick a routine you can stick with, like 5–10 minutes of practice before dinner. Try finding ways to work training into your daily routine, like having your pup practice sitting before you put down their food bowl, doing a sit-stay in the elevator, or walking nicely on the way to the dog park.

GET USED TO WORKING FOR CRUMBS

Realizing I got snackies for doing what the human asked was life-changing, but at first I questioned the absurdly small size of each one. I knew my worth. It was way more than crumbs. But one night I found a secret stash of treats and ate a few more than I should have. And I can't believe I'm saying this, but … it's technically possible and regrettable to eat too many snackies. Think extra late-night trips outside, not to mention outgrowing your turtleneck sweaters. So yeah, actually, maybe be grateful you only get a crumb each time.

Training requires a lot of practice before commands become second nature. Your pup will sit again and again (and again), so give them a very small treat for each rep to avoid stomachaches or weight gain, and eventually wean your dog off getting a treat after every repetition of a command. What's "small" depends on size—big dogs can get more than a literal crumb, but still be mindful that less is more. Side note: make sure you don't leave treats, or anything else potentially harmful where your dog can find them.

WORK HARD-ISH, PLAY HARDER

Remember when I said rules apply outside? Well, guess what … rules also apply *during playtime*. Yeah, I know. This was irritating at first, but then I started pretending that doing what the human asked was part of the game. Like some secret cheat code I have to enter before she releases the toy. A good pre-throw "stay" really gets the adrenaline pumping, all that waiting and anticipating the throw so you can run out and attack that squeaky bunny like the ferocious lil' pupper you are.

Playtime is great for training. For many dogs, a toy can be just as valuable a reward as a treat. Ask your pup to sit, stay, or whatever else you're practicing before throwing the ball. If your pup is overexcited, practice manners and patience by waiting to resume the game until your pup "asks nicely," i.e., offers a quiet sit or lies down. Think of these things as the puppy equivalent of zucchini in brownies: secretly sprinkled in for good health while still keeping things mostly yummy and fun.

A LOVE LETTER TO TRAINING

Sproutie was barely 2 pounds (900 grams) when he came home, and the world must have felt overwhelmingly big and confusing to him. I mean, he was *scared* of Velcro®. But as soon as he started classes and figuring out what he was supposed to do, his confidence grew. Watching tiny Sprout come out of his tiny shell through training showed me that he was not only incredibly smart but, more importantly, happy.

We learned so many valuable lessons beyond the basic commands. I gained tools and advice on how to actually raise a well-adjusted puppy. Sprout learned how to play the game of training (more from him on that later), and I learned the training principles necessary to keep working with him on my own. I also learned to give up on perfection. I am not perfect. And—I'm covering his ears as I write this, but—Sprout is not perfect. We make mistakes and at times fumble through training. Sometimes I ignore my own advice and, well, on occasion he ignores me. But with our training foundations, and our efforts to do it right almost all the time, we have succeeded.

Best of all, we bonded. Sprout was actually fairly indifferent to me at first. Sure, he'd greet me when I came home and follow me into the kitchen in the hopes I'd drop something, but it wasn't until training that we really connected and grew into a team.

So thank you, dear training and trainers, for all you did for Sproutie and me. We love you.

LOVE THE CLICK

Only one sound rivals the telltale crinkle of the snackies bag—the magical "click." The sound may seem random, perhaps off-putting, at first, but soon you'll appreciate all the promise it holds. See, a "click" means you did good-dog and you're about to get a snackie! Whatever you were doing right as the human clicked is what they flagged as the good-dog action, so just keep repeating that.

If you're like I was, you may be clueless about what "clicker training" is, so here's your clue: it's a reliable way to tell your dog they did the right thing and will be rewarded. Clicking exactly as your dog does what is asked is called "marking" the behavior. Saying "good" or "yes" works as a marker too, but our voices aren't consistent—"GOOD!," "Good boiiiii!," and "Gooooooddd" sound like different things to your pup. The "click," however, always sounds the same. Start by repeatedly clicking and tossing your pup a treat to help them recognize and understand the click. Clicker training is most useful when learning something new, but you can phase out using it as your pup progresses.

GOOD THINGS COME TO
THOSE WHO LISTEN

At first, I was quite disgruntled by my human's commands. I mean, they weren't crazy asks, and I was generally inclined to lie down anyways, but the idea that she could dictate what I did seemed inherently unfair. Sure, suggest away lady, but *command*? But then I realized ignoring my human meant missed opportunities for snackies. And that seemed insane. So yeah, I'm a devoted listener now.

When you give a command, you want your pup to follow it immediately. Sounds simple, but we've all watched someone say, "Sit! … Siiiitt … Ugh, SIT" to a dog who just indifferently smiles back. Repeating yourself implies listening is optional—if you're cool with your pup not sitting until the third command, then were the first two just suggestions? Try getting their attention before even giving a command. Your pup's more likely to obey if they're already listening, helping create a habit of quick responses. Practice holding your dog's attention by rewarding them for making eye contact with you.

COME HERE, OFTEN

I have many gripes, such as the human's insistence upon waking at a respectable hour, and the fact that she makes it do the wets outside. However, I do enjoy her company and find it necessary to be attached to her side at all times no matter how inconvenient she may find it. Needless to say, I chuckled to myself when I realized she was trying to teach me to actually come closer to her. Seriously lady, you think you need to give me snackies for lying down *and* for running over to you? Lol, gotta love human logic.

One of the most useful things I learned in puppy class was to get Sprout to touch his nose to the palm of my hand. This "hand touch" became the basis for "come" by moving away so that Sprout needed to come to me to reach my hand. Practice this behavior by playing a super cute game of "puppy ping-pong": have a family member or friend stand across the room from you, and take turns cuing your pup with a hand touch, getting them to excitedly run back and forth.

YOU'RE NEVER TOO OLD
FOR NEW TRICKIES

Age is just a number and considering you can't actually count numbers, age really doesn't matter. So don't ever let anyone tell you that you're too old to learn something new or to change. Granted, you may be more stubborn and not see the need to interrupt your nap with silly activities, but you certainly *can* learn something. Taking up a new hobby later in life can be surprisingly rewarding, so allow me to encourage you to do so.

Even adult dogs who know the basics can benefit from taking a class or learning a new trick. Training is a great way to keep your dog's mind engaged and add a little fun to their daily routine. Think about your pup's personality and find an activity that suits them. Whether you try agility classes, nose work, dock diving, or even just a good ole-fashioned trick class, the options for you and your dog are endless.

PLAYING THE GAME

Listen up puppers because I'm about to teach you a major life lesson: learning is not a lackluster chore, it is actually the goodest game. EVER. You are playing with your human *while getting snackies*. Stop and let that sink in for a moment. Can you think of anything better? I can't. Okay, actually, maybe being fed while getting a soothing little belly-rubbies is better. But at any rate, back to this game.

It goes like this: figure out what the human wants and do that as fast as you can to get a snackie. Don't just guess—actually focusing on what the human wants is a better strategy in the long run. Your goal is to complete as many levels as you can. And here's the cool part: the game gets harder as you level up, so you're always strategizing. Level 1 may just be touching a human's hand with your nose, but by Level 8 you have to wait still until you're released and then sprint across the room and down the hall to find and touch the human's hand. When you beat a level you get this rush that makes you keep coming back and striving for more. So go on, don't just play the training game—win it.

CHAPTER 5

EXTRACURRICULARS

THAT TIME I ALMOST JOINED THE CIRCUS

A few months into school, I was proverbially prancing toward life as a scholar. I mean, the human still had to carry me up the stairs, but now I sat front and center in class, titling my head as I took copious mental notes. It was the first day of a graduate-level class called "Circus Tricks," and my professor, Ms. Joanne, said we'd cover a lot of trickies and that some would be easier than others, so we should each just go at our own pace. And then she added, "By the end of the course, I bet one of you pups will be able to do a handstand!" I jumped to my feetsies in excitement and looked back at my human. She nodded and winked. It was in that moment I knew I'd found my true calling. I would do the impossible. I would learn to do a handstand *without even having hands.*

I embarked on my mission with rigorous training. I did puppy push-ups and practiced kicking my hindlegs up higher and higher. I ran laps around the coffee table to develop endurance, and meditated to maintain focus. Several weeks later, armed with nothing more than feetsies and the mental fortitude of a champion, I found myself in a handstand. My world was forever changed. Turned upside down, if you will.

I was so excited for graduation that I bounded up the first two stairs to class before realizing I needed human assistance. Eventually, it was time for my graduation performance. I wove between my human's legs as she walked. I jumped through her arms. I did countless silly acts all in the name of snackies. And at the end I popped up into a handstand. Humans applauded. A sheepdog looked away sheepishly. Ms. Joanne smiled and gave me my diploma. I was glowing.

The afterglow, however, faded fast. Later at home, I glimpsed something on the countertop. It could have been nothing, but, more

importantly, it could have been a bag of snackies. I made an inquisitive whimper and looked back at the human. She sat there doing whatever it is humans do. No reaction. I let out a determined grumble and stared very hard, first at the countertop, and then back at the human. No movement. I was growing quite irritated. Perhaps even infuriated. I was entitled to know what sat high atop my countertop. Grumblings turned to yips and then to fully disgruntled yaps. Still no reaction.

I couldn't believe it. After all I accomplished, all the snuggles I gave, and the rewards I so graciously received, I was ignored? Intolerable. I grabbed my favorite ball and marched to the door, determined to make it on my own. I would join the circus and show off my handless handstand. I would be adored, not ignored. I didn't need my human.

But then I realized I couldn't open the door without her. I sat and stared up at the handle, plotting my next move. "Aww, good boy, do you need a walk?" she cooed as she stood up. Seriously? *Now* she noticed me. Wait, yes, this was it—I'd let her walk me like a good boy and then I'd make a break for it and hitch a ride to the circus.

It was cold and dark outside. I looked up at the human. She smiled and bent down to give me a chin-scratchie. My insides started to warm. Wait, no, I could not fall for this. *Be adored, not ignored*, I reminded myself. I took a deep breath and bolted. Life flashed before my eyes. That insensitive frog. Snuggles. The wets. Snackies. Fetchies. Stairs. Chest rubbies. Being *ignored*. Being *adored*. Emotions poured out as I ran for my life, but four steps later I snapped back to reality.

It was the end of the leash. Oh, right. I couldn't just run away with the human tethered to me. "What was that about, Sproutie?!" she said, scooping me up and tucking me inside her coat. My insides started to warm, and this time I allowed it. I mean, maybe I *was* adored. Maybe life wasn't so bad with her tied to me.

JUST BE YOURSELF

By now you must believe I'm perfect. I mean, you *are* talking to the pup who nailed the handless handstand. But, the reality is, I have shortcomings. No, we are not talking about my height. I can't do one of the oldest trickies in the book: roll over. I get on my side but then I just flail, and the awkwardness isn't worth a snackie. But I've realized I don't need to roll over to lead a fulfilling life. I just need to be me, and that's about as perfect as perfect can be.

All dogs are undeniably perfect, but they're not created equally. One look at Sprout's posture and you'll see he's a rigid dog. He doesn't like being on his back, and while he's smart enough to learn to roll over, he wouldn't find it all that enjoyable, so we skipped it. Pick tricks that your dog will enjoy learning by considering their natural strengths and tendencies—while recognizing their particular physical limitations. Here are a few of our favorite simple tricks to start you on your way!

LEARN "BOOP!"

I don't know about your human, but mine used to do this thing where she touched my nose with her finger while saying, "Boop!" in her silly voice. It was endearing, but also kinda just rude. I'm a person too and should have a say over whether or not there will be a nose boop. So we reached an understanding wherein she puts her finger out and asks to do a boop! and then I decide whether or not to engage in the boop. And I always do because, to be honest, it's pretty fun.

As I was sitting there booping Sprout's little nose, I wondered whether it was annoying. He didn't seem to mind it, but I figured why risk it, and so I turned it into a trick. It's a similar concept to the hand touch, but cuter. If your pup is a little mouthy or nips, you may want to skip this trick or proceed cautiously to avoid an over-enthusiastic nibble on your finger!

Turn the page to discover how to learn and perfect this trick …

The Set-Up: Stick your index finger straight out like you're pointing. Hold your fingertip near your pup's nose and give out the cue, "Boop!"

The Idea: Your pup should lean in to sniff your finger, and right as their nose touches your finger, "mark" what they did (click or say "good") and give them a treat. Most dogs will want to sniff what's in front of them, but if yours plays hard to get, try luring them in with an enticing food scent or tiny dot of peanut butter on your fingertip.

The Practice: Keep repeating this. At first, mark and treat anything that comes close to the desired result, like their nose against your knuckle. Right now, you want your pup to get the general idea that seeing your finger means they come touch it with their nose.

The Perfection: Now it's time to refine the trick. Only mark and treat when your dog's nose actually touches the very tip of your finger. No treats for knuckle-boops. Being selective about which attempts get treats and which don't will help your pup learn what exactly they are supposed to do. Like with most tricks, once your pup is doing the behavior consistently you can start phasing out treats, like rewarding every third or fourth boop!

LEARN "ROOM SERVICE"

I stand behind my debatably undebatable claim that an old-fashioned sit is 21.3 times more effective at getting snackies than barking. But I have an indulgent little secret: there is a device that, when tapped with your paw, makes a little yippy ring sound which causes a human to present you with a snackie. Like, without fail, if you tap it, a morsel appears. If a human is seated and you tap it, they must rise and fetch you a snackie. I highly recommend sneaking the device into your bed, snuggling in, and ringing away to your heart's content.

Sprout's correct that every time he rings the bell he gets a treat—tricks are easier for your dog to learn when the rules are consistently applied, so bell always equals treat. Make sure to store the bell out of your pup's reach when not in use, otherwise you risk creating a little bellringing fanatic. If your pup is very sensitive to noise, it's best to skip this trick or proceed cautiously as you get them accustomed to the ring.

Turn the page to discover how to learn and perfect this trick …

The Set-Up: Find a bell that'll be easy for your pup to push down on—like the type you see on a service desk. Push buttons work as well. Help your pup start associating the bell sound with treats by spending a few minutes tapping the bell yourself and tossing them a free treat with each ring.

The Idea: To get your pup to tap the bell themselves, place a tiny treat under the bell while your pup is watching. Odds are they'll come over, sniff a bit, and then paw at the bell to get the treat. (You may need to hold the bell in place so they don't knock it away.) When they make the bell ring, pick it up so they can get the treat.

The Practice: Keep repeating with a treat under the bell. When your pup is doing this consistently, put the bell down but keep the treat in your hand. Give your cue (try "room service" or "ding!") and when your pup taps the bell, toss them the treat.

The Perfection: Once your pup is tapping the bell on cue, refine the trick by adding some distance. Start by placing it a few inches out of their reach and gradually increase the distance until your dog is running across the room to tap the bell. Feel free to get creative—for example, have your pup ring the bell before being served dinner.

LEARN "STRIKE A POSE"

Humans seem obsessed with documenting the lives of their puppers. Like, I know we're all unbelievably cute, but must you capture every moment? And what are you even doing with all those pictures? *Does anyone even see them?!?* Anyways, I don't mind posing. The human asks me to do something, like sit on a chair or whatnot, and I decide if I do it. And I always do, because putting my feetsies up on the furniture is fun, and then I make it look all sassy and everyone gushes about how perfect I am and gives me snackies. So yeah, document away!

We all love taking pictures of our pups, but it's so important to make sure they love it too. Training your dog to pose is a great way to make them an active participant, and helps ensure they feel comfortable and confident during a shoot. This two-paws-up trick is an easy and versatile pose to start with.

Turn the page to discover how to learn and perfect this trick …

The Set-Up: Find a box or container that's no more than half as high as your pup's legs and sturdy enough to support their weight. Position your pup so they're standing facing the box and you can "lure" their front half up onto it.

The Idea: Hold a treat near your pup's nose and slowly move it away from them so it's above the box—you want your pup to lean in to follow the treat. As you move the treat away, they should naturally put their front paws up on the box to reach closer. When their front paws are up on the box, "mark" it with your clicker, or say "good," and give them the treat. If your dog isn't putting their front paws up on the box, try luring them onto a lower box or large book to start with. Keep repeating this process until your dog seems used to this motion.

The Practice: Once your pup has the general idea, repeat the "luring" motion, but with only your hand and no treat in it. Still "mark" and treat as soon as their paws are on the box. Eventually start minimizing the use of your hand by tapping the top of the box instead, and encouraging their front paws onto it. Continue marking and rewarding as before. Your goal is for your pup to eventually pop their front paws up when you simply tap or point to the box.

The Perfection: Once you've gotten them popping up onto the box regularly, start practicing with different boxes, heights, furniture, objects, etc. Pretty soon they should realize the trick applies to any object you tap and cue. Make sure you're using stable objects and surfaces to avoid any slips, and take particular care if you progress to posing at a table. Both the chair and table need to be stable, and the chair should be close and high enough that they can easily and safely reach the table. If you tap the table and they don't "strike a pose," this could be a sign they aren't comfortable, so step back and rethink your plans.

LEARN "DANCE PARTY"

A dance party is where humans jump around and be silly and have fun. It is considered to be 6.2 times more fun if a puppers attends the dance party, and 11.4 times more fun if the pupper is also able to dance. Dancing, fortunately, is a fairly easy human behavior to learn—you just basically stand up and, well, act like a silly human. Pretty fun opportunity if you ask me.

As a very small yet very curious puppy, I noticed Sprout often popped up onto his hindlegs to try and see what was on a table or countertop, so I decided to use that natural behavior for a trick. We took it to the next level by making the cue a fist pump, so if anyone is dancing around while doing a fist pump, Sprout will jump up and make it a dance party. This trick is more physically engaging, so if your pup has a long body and/or potentially bad back or weak hind end, check with your vet before trying this trick.

Turn the page to discover how to learn and perfect this trick …

The Set-Up: No props needed for this one, but you must be on carpet or another safe surface where your pup won't slip when they jump up.

The Idea: Start by giving your pup a few free treats while they are standing in front of you to get them excited. Hold the next treat above them, an inch or two out of reach, so they have to jump up a little to reach. Give them the treat while their front paws are in the air. Don't worry about them being upright on two legs yet, just keep practicing until they're used to jumping up slightly.

The Practice: Gradually start holding the treat higher and higher, continuing to mark and reward while their front paws are up, until eventually they are upright. Keep practicing upright for a bit. To help them start staying up on their hind legs longer, delay the reward by holding onto the treat a second or two before giving it to them.

The Perfection: Just dance around the living room with your pup like no one's watching and HAVE FUN!

CONCLUSION

A STAR IS UNLEASHED

Two minutes until the Sprout half-time show … ARE YOU READY?!?!

The announcer's voice echoes throughout the large room of humans. I'm about to take the stage at my third PetCon™ event, and performing this gig is way better than being in the circus. The human grabs my shopping cart and props, checks her cue cards one last time, and takes a deep breath. I stretch, shake it out, and sneeze loudly with delight. "Oh, are you ready, puppers?" she asks. I high-five her like there's no tomorrow. "Ready, ready, READY?" she shouts in her ridiculously excited voice. A high-ten this time. *I was born ready, lady. Let's do this.*

I prance my way to center stage. Cheers, claps, and I'm pretty sure a standing ovation before we even start. "Hey Sprout-Sprouts, can you say hi?!" asks my human. I look out into the audience and wave hello. They're smitten after one trickie.

We keep going, me showing off every last bit of good-boy I know. I'm up on my hind legs dancing like everyone is watching. Because they are. She tells funny stories about how magical and wonderful I am.

Humans laugh. Like, *with* her, not at her (I think). She explains important things like how to use training to make sure your pup stays happy and confident, and your pup decides whether to boop! or not. And I realize: I am the embodiment of everything she is saying. Wow, that's like, really deep.

But there's no time to reflect, only to shine. Adrenaline is pumping, and I'm so focused that, for a moment, I forget we aren't at home practicing trickies before my pre-dinner nap. But as I drop a banana into my shopping cart with expert precision, I catch a glimpse of the crowd.

They cheer wildly as I waddle across the stage pushing my cart. I beam with pride, thinking back to when we fumbled our way through learning "Down." When I was terrified of Velcro® and doorways. Nothing more than a tiny puppy lost in the great big world, trying to train a

human while training myself. But look at us now. Me, the natural-born performer, and her, the ever-faithful stagehand. The unstoppable duo. We did it, we conquered the world. Just me, her, and the snackies.

IMPORTANT NOTES
FOR HUMANS

I have two goals: to make you smile at our bond with dogs, and to provide real, relatable advice. The validity of our guidance is deeply important. I've looked to those with professional dog and veterinary training to help review this book. I took some creative liberties with Sprout's thoughts, but encourage you to learn how your dog actually perceives the world. We're excited for you to start your training journey! Here are some resources we consulted and think will help you along the way:

Understanding your dog's thinking:

American College of Veterinary Behaviorists, 2015. *Decoding Your Dog: Explaining Common Dog Behaviors and How to Prevent or Change Unwanted Ones*. Boston: Mariner Books, www.dacvb.org

Coren, Stanley, 2004. *How Dogs Think: What the World Looks Like to Them and Why They Act the Way They Do*. New York: Simon & Schuster.

Horowitz, Alexandra, 2009. *Inside of a Dog: What Dogs See, Smell, and Know*. New York: Scribner.

Positive reinforcement methods:

Arden, Andrea, 2007. *Dog-Friendly Dog Training*. 2nd Ed. New York: Howell Book House.

The Monks of New Skete, 2011. *The Art of Raising a Puppy*. Revised Ed. New York: Little, Brown and Company.

Stilwell, Victoria. "Positively®.": www.positively.com

Crate training:
Garrett, Susan. "Crate Games for Self-Control and Motivation.":
   get.crategames.com

Clicker training:
Pryor, Karen. "Clicker Training.": www.clickertraining.com

Learning tricks:
Sundance, Kyra and Chalcy, 2007. *101 Dog Tricks.*
   Beverly, MA: Quarry Books

Finding a dog trainer:
The Association of Professional Dog Trainers: apdt.com
International Association of Canine Professionals:
   www.canineprofessionals.com

Canine sports and activities:
American Kennel Club Sports & Events: www.akc.org/sports
The Kennel Club (UK):
   www.thekennelclub.org.uk/events-and-activities

Animal welfare:
American Society for the Prevention of Cruelty to Animals:
   www.aspca.org
The Humane Society of the United States: www.humanesociety.org
National Brussels Griffon Rescue, Inc.:
   www.nationalbrusselsgriffonrescue.org
Royal Society for the Prevention of Cruelty to Animals:
   rspca.org.uk

ABOUT SPROUT & SIGRID

Sprout, the star of the critically-acclaimed Instagram account @brussels.sprout, is a perpetually-pouty, smooth-coated Brussels Griffon with a repertoire of over 30 tricks and cued behaviors. He enjoys long walkies on the beach, napping in the sun, and tennis balls. He has no idea what Instagram is.

Sprout owns human Sigrid. A lawyer by day, @brussels.sprout is the creative outlet she never knew she needed, combining her life-long love of animals, learning, and wit, to inspire people and pups to bond through training. Sprout and Sigrid live in Manhattan, where you can find them playing fetchies in the park.

ACKNOWLEDGMENTS

Thank you, thank you, thank you, to …

Mom and dad, for your rules of life, writing genes, and endless support. Tins, for always being someone to look up to. Abigail, for being the best ball thrower ever. Loni, for encouraging me to write this book (and always being right). Jessica, Laura and the White Lion team, for believing in our vision and giving it life.

Our Brussels Griffon friends, for welcoming us into your world. The @brussels.sprout community, for letting us make you smile. Everyone who's ever told Sprout he was a good boy. Chelsea Wine Cellar, for all the snackies bestowed upon Sproutie.

Those who dedicate themselves to training, understanding, and protecting dogs—we wouldn't be here without you. Niki, for your support on and off the agility course. Andrea, for helping us fall in love with tricks. Joanne, for the handstand.

And thank you, Sproutiekins. Somehow, I end this book at a complete loss for words to describe how much you mean to me, but I hope every chest-rubbie, fetchies game, sunny-spot relocation, and free snackie lets you know. I also hope that one day, you realize I don't control the weather, but I digress. I love you.